# BUILDING STIRLING 1
## A One Piston Hot Air Engine

by

E. T. Warbrooke

British Library Cataloguing-in-Publication-Data: a catalogue record of this book is held by the British Library.

First Printing 2002

ISBN No. 0-9536523-4-3

Published in Great Britain by:

**CAMDEN MINIATURE STEAM SERVICES**
Barrow Farm, Rode, Frome, Somerset. BA11 6PS

Camden stock one of the widest selections of engineering, technical and transportation books to be found; Write to the above address for a copy of their latest free Booklist.

Layout and Design © Camden Miniature Steam Services
Layout and Design by Camden Studios & Andrew Luckhurst, Trowbridge, Wilts.

Printed and Bound by Salisbury Printing Co. Ltd.

**Publishers Note:**

The machining operations, and running of the engine, described in this book can potentially be hazardous. The author and publisher are just passing on information -
Safety is entirely your responsibility.

# CONTENTS

## ABOUT THE AUTHOR:

Ted Warbrooke describes himself as "semi-reitired and 60 plus". As a youth he lived near to New Zealand's great boating harbour in Auckland, and was fascinated by the many working engines being used around Auckland at that time.

He travelled New Zealand and Australia for many years as a mechanical fitter and came to believe that the simple ideas that work well are the best.

During the 1990s the author learned how to make Stirling engines from some old copies of *Model Engineer* magazine, but soon vowed to simplify the Stirling engine, as there semed to be no simple heat engines a beginner could make.

To date the author has made dozens of smll heat engines, and has also come to believe that almost any sort of heat engine will run, as long as the moving parts are made light-weight and very free moving.

Some of the author's other interests include Science, Electronics and, in general, how the things of this world 'tick'.

## PUBLISHER'S NOTE:

When Ted Warbrooke first sent me the draft of this book I have to admit that I was sceptical as to whether a "displacerless" hot air engine could work, a scepticism that was finally overcome when Ted sent me one of his engines.

In the interim I had discussed *Stirling 1* with Bob Sier, well known guru on the Stirling Engine, who confirmed that he had heard about such an engine, and subsequent research by him revealed a number of patents for hot air engines without displacers dating back to the 1950s, some intended as heat pumps rather than engines. Whether any such engines were built is currently not known and the fact that they were referred to as either Lamina Flow engines, or acoustical heat engines, suggests some confusion as to how they operate. One page 21 of the present work Ted Warbrooke suggests how his *Stirling 1* engine works, but no more than that.

From the same page, the reader will gather that Ted arrived at *Stirling 1* by another route, and his own intellectual capabilities, and there is no doubt that *Stirling 1* works, and works well.

*Stirling 1* represents the simplest form of prime mover there is, and it is difficult to imagine one that could be simpler. As such it is an ideal project for the beginner to model engineering, but it also offers considerable scope for experimentation to the more advanced hot air engine builder. And if you can explain the exact thermodynamics we will be pleased to hear from you.

Adam Harris
(with acknowledgments and thanks to Bob Sier)

Camden Miniature Steam Services

# INTRODUCTION

**SINGLE PISTON - CLOSED CYCLE**

Just the one piston works this Hot Air Engine rather than the usual two pistons. Because the power of this engine is low it MUST be made very free moving to run well. The air in the engine is used over and over again. *STIRLING 1* will run well in either direction.

**A SIMPLE CYCLE**

Due to the thermal action within the heater the 'WORKING PRESSURE' falls during the compression stroke then rises during the hotter expansion stroke of the piston, so driving the engine. The heat of a small spirit flame is all that is required to run the engine.

**DOUBLE-ACTION**

Keen experimenters may wish to make this type of engine double-acting by driving both faces of the single piston.

**PRE-RUN TEST**

When the engine is 'FLIPPED OVER' by hand to give it compression it must have a very bouncy feel to it. There should be no binding or tight spots when the crankshaft is rotated.

**WORK TOOLS**

At least a small metal-working lathe and a bench drill along with the usual hand tools are needed to make the parts of the engine plus ideally access to a small gas welding set, or TIG welder to complete the heater.

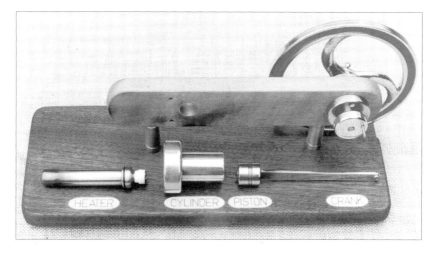

Components of Stirling 1 - note hole in centre of body for a side mounted heater

# *STIRLING 1* ENGINE
General Layout

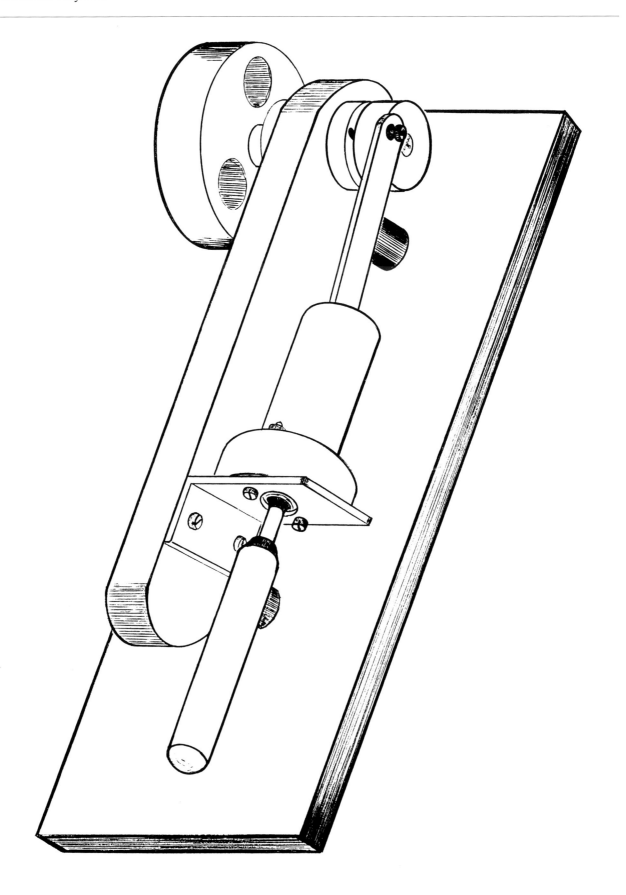

# STIRLING 1
One Piston Hot Air Engine: Working Principle

## CUTAWAY VIEW
Heater unit, Cylinder, Piston

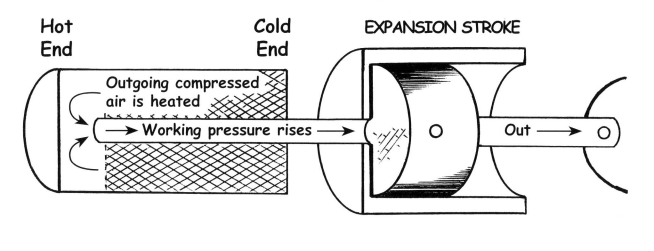

**Hot End**    **Cold End**    **EXPANSION STROKE**

Outgoing compressed air is heated

→ Working pressure rises →

Out →

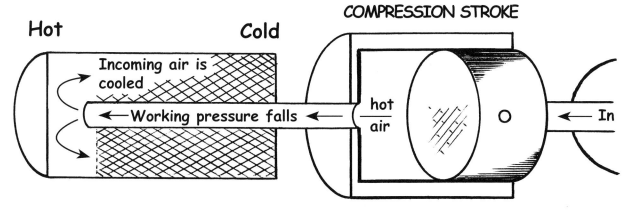

**COMPRESSION STROKE**

**Hot**    **Cold**

Incoming air is cooled

← Working pressure falls ←

hot air

← In

A simple, one piston, Stirling Cycle Engine which will run in both directions.

Using both faces of the power piston the engine can be made double-acting.

**NOTE**
The types of fasteners, and the threads, shown for this design are those the author used in making *Stirling 1*. They are not set in stone and the reader can vary these to suit himself, although similar thread sizes should be used.

# HEATER/REGENERATOR UNIT

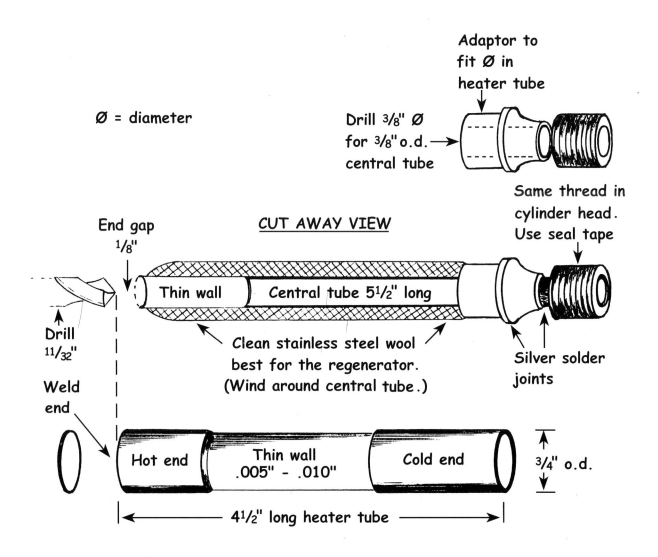

Ø = diameter

Adaptor to fit Ø in heater tube

Drill 3/8" Ø for 3/8" o.d. central tube →

Same thread in cylinder head. Use seal tape

End gap 1/8"

CUT AWAY VIEW

Thin wall    Central tube 5 1/2" long

Clean stainless steel wool best for the regenerator. (Wind around central tube.)

Drill 11/32"

Weld end

Silver solder joints

Hot end    Thin wall .005" - .010"    Cold end

3/4" o.d.

|← 4 1/2" long heater tube →|

## POSSIBLE MATERIALS FOR HEATER UNIT

Stainless Steel, Steel, Bronze

- Outer heater tube and end plate: Thin-wall Stainless Steel - a suitable sized glass test tube could also be used.

- Central air tube: Thin-wall Stainless Steel.

- Regenerator material: Stainless Steel Wool (pot cleaner), thin metal shim, loose wound silver paper.

- Adaptor: ¾" round solid mild steel.

- Threaded end: ⅝" fine thread bolt end.

All the parts must be thoroughly cleaned before the heater is assembled.

**Stirling 1 Heater Unit to the design on the lefthand page**

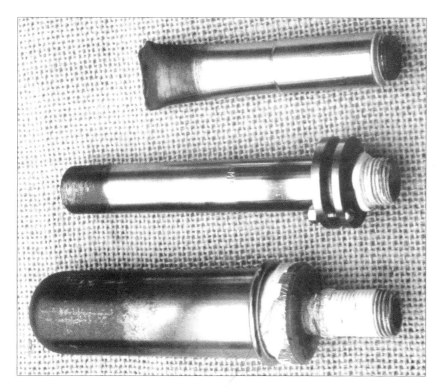

**Three Different Complete Experimental Heater Units**

If possible make the heater's outer cover from a good grade of Stainless Steel. This will reduce metal scale problems. A good quality glass test tube, if one can be found of the appropriate dimensions, can also be used but extra care must be used in fitting it.

Stainless Steel is sometimes hard to gas weld so TIG welding may be needed to complete the 'hot end' of the heater unit. Silver soldered joints at the 'cold end' are often no problem.

Stainless Steel Wool (pot cleaner) makes a good, long lasting regenerator but other thin, non-rusting materials often work quite well.

Simply made heaters with crimped 'hot ends' and glued 'cold end' fittings, as shown on the left, have been used but all the joints <u>must</u> be kept airtight. *STIRLING 1* has been worked with several types of heater; however, the heater unit that best suits this engine is shown with these plans and in the photograph above.

## ENGINE CYLINDER
Brass or Manganese Bronze

---

**HELP FOR BEGINNERS:**

- Take care when using emery paper around the lathe - in particular always cover up the bearing surfaces.

- If possible avoid using Aluminium Bronze, this metal is often hard to machine and hard to soft solder. If used, *Loctite* may be a suitable alternative to soft soldering.

- The Engine Cylinder is made from a piece of extruded brass or bronze tubing. This stock usually has a very smooth bore and is ideal for an engine cylinder because of its accuracy.

- When you machine brass, and some of the bronzes, your cutting tools may tend to 'dig in'. To prevent this use flat top cutting tools in the lathe. Twist drills should be de-raked or straight fluted when cutting these metals.

- Finish making the engine cylinder before you make the piston, you will find it easier this way than to make the piston first.

- Don't distort the engine cylinder by holding it too tightly in the lathe chuck. Face off both ends of the cylinder in the lathe and round off the sharp corners.

- Smooth the rounded corners of the cylinder with fine emery paper, don't use the emery paper down the bore of the cylinder if the cylinder bore looks smooth. Clean away all emery dust.

- Polish the bore of the engine cylinder with a wooden lap charged with *Brasso* and light oil as shown on Page 7.

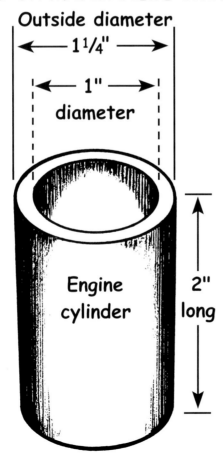

### ENGINE CYLINDER
#### BRASS OR MANGANESE BRONZE
Outside diameter
1¼"
1" diameter
Engine cylinder
2" long

## CYLINDER POLISHING LAP
Soft wooden dowel

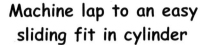

Machine lap to an easy
sliding fit in cylinder

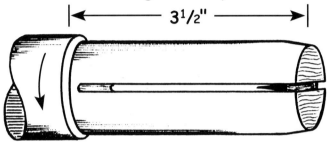

CAUTION: Cover up lathe bearing
surfaces. Clean up the cylinder
and lap afterwards

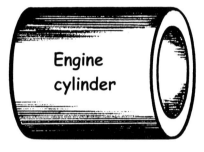

To and fro along
the lap by hand

### REDUCE THE FRICTION IN YOUR ENGINE - POLISH THE PARTS

When you have finished the outside diameter of the wooden lap to size, lightly oil the surface of the lap and add to this a few drops of *Brasso*.

Revolve the lap at a slow speed and move the engine cylinder to and fro along the length of the lap by hand until the cylinder bore is polished. Sometimes only a dozen or so strokes are needed to complete this task.

If the wooden lap becomes worn etc. split the lap up the centre with a hacksaw cut and use it as an expanding lap. It is not a good idea to use the engine piston as a cylinder lap if you are a beginner. Serious damage could result to these parts.

There is no need to polish out every tiny mark on the bearing surfaces; remember that you are not making a watch here. Polished bearing surfaces run with the right clearance need no running-in time.

# ENGINE PISTON
Cast Iron, Mild Steel or Aluminium

---

## HELP FOR BEGINNERS

- Prevent the polishing agents or the emery grit from entering between the moving surfaces of your machines. Cover these up.

- Machine the outside diameter of the piston to slightly larger than the cylinder bore (.001″ or so).

- Round off the sharp corners at both ends of the piston with a fine cut flat file.

- File the outside diameter of the piston with care. When the piston will just enter the engine cylinder without forcing it, finish with fine emery paper (1200 -1500 grit).

- Protect this fine finish on the outside of the piston. The piston should be a neat sliding fit in the engine cylinder.

- Drill out the inside of the piston to take the gudgeon block etc. before parting piston off.

- Don't deform the piston walls when machining the piston crown.

- Slightly countersink the gudgeon screw hole, thread tape sealer is used here later.

- The bearing surfaces of the piston and cylinder MUST be kept clean, polished and lightly oiled if the engine is to run well. Pistonhead - air-gap should be about $1/16″$.

- The piston skirt can be increased in length to aid compression if neccesary, but by no more than $1/2″$.

- Two shallow grooves made in the skirt of the piston will aid lubrication

### PISTON CUT AWAY VIEW

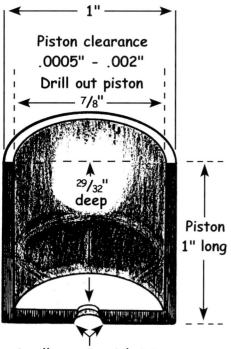

1"

Piston clearance .0005" - .002"
Drill out piston
7/8"

29/32" deep

Piston 1" long

Small countersink into 5/32" Gudgeon screw hole for sealer tape

### SAFETY:

**Take care when hand finishing and polishing the piston in the lathe**

Round bar

## ENGINE CYLINDER HEAD
Brass or Bronze

---

### ENGINE CYLINDER HEAD

#### BRASS OR BRONZE

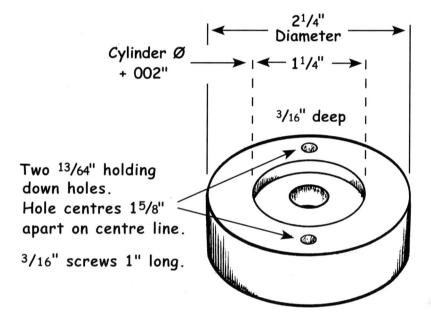

2¼"
Diameter

Cylinder Ø
+ 002"

1¼"

3/16" deep

Two 13/64" holding
down holes.
Hole centres 1⁵/8"
apart on centre line.

3/16" screws 1" long.

### CROSS SECTION VIEW

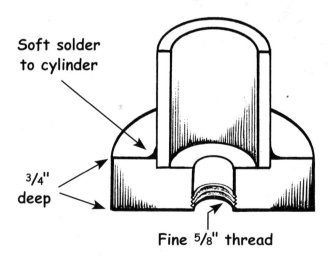

Soft solder
to cylinder

3/4"
deep

Fine ⁵/8" thread

This head design is for an in-line heater unit, but the heater can be fitted to the head in almost any position as will be seen from the pictures later in this book.

- Make the cylinder head of Brass or Bronze but not of Aluminium Bronze.

- Both faces of the cylinder head should be turned in the lathe and be parallel to each other.

- Make a short, fine pitch ⅝" thread in the base of the cylinder head to take the thread of the heater tube. Short threads slow heat transfer to the head.

- Soft solder the cylinder unit together. Make sure that the surfaces to be soldered are clean metal, not tarnished. Run a small airtight fillet of electrical solder around the cylinder/head joint when the unit is hot enough to melt the solder wire. You could possibly use the heat of a stove element or the heat of a soft gas flame to do the job.

- Ensure that there are NO Air-leaks where the heater unit is screwed cylinder head - the engine will not run if there are any air leaks.

## PISTON GUDGEON BLOCK
Aluminium or Brass

### CROSS SECTION VIEW

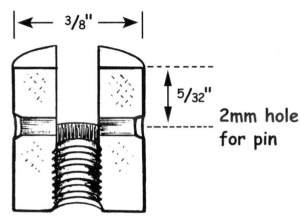

Tap 5/32" Whitworth

### CONNECTING ROD SLOT

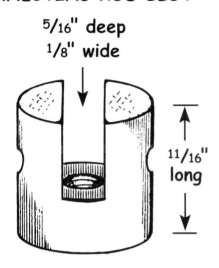

## PISTON/GUDGEON BLOCK ASSEMBLY

- There are many ways of making the Gudgeon Block but whichever way you choose to make it, the gudgeon or wrist pin hole should always be at an angle of 90 degrees to the sides of the piston.

- The engine piston should be made to swing easily from side to side when it is attached to the end of the connecting rod.

- Small ball joint units can sometimes be found. If you are able to use a ball joint unit here instead of a home-made gudgeon unit it may work better and save you time.

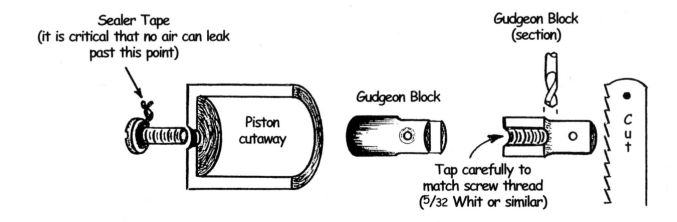

# CONNECTING ROD
Aluminium Flat Stock • ⅜″ wide • ⅛″ thick

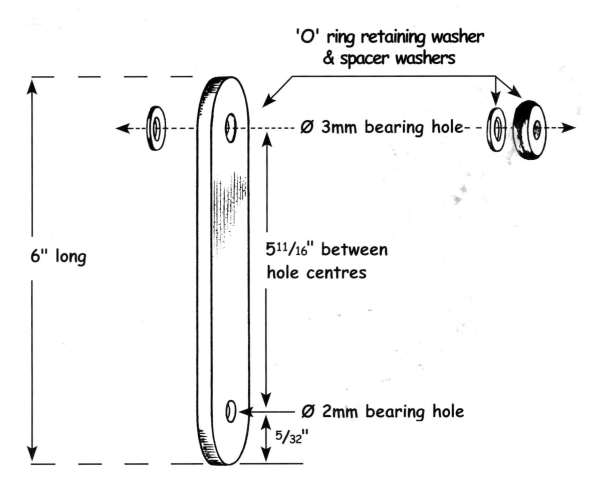

'O' ring retaining washer & spacer washers

Ø 3mm bearing hole

6" long

5¹¹/₁₆" between hole centres

Ø 2mm bearing hole

5/32"

- The crank and gudgeon pins should be an easy running fit in their respective bearing holes in the connecting rod.

- Drill several holes in a piece of metal off-cut to get the correct fit for the pins then drill the 2mm and 3mm diameter holes in the connecting rod.

- Feed the connecting rod under the drilling head in one direction only. The connecting rod bearing holes will be in line with one another by doing this.

- Slightly countersink both holes on both sides of the connecting rod; you will find the pins enter their respective holes more easily if this is done.

- When running the engine, keep the crank-pin well oiled, or fit a ball-race.

## CRANK AND GUDGEON PINS
Stock Needle Rollers

---

# CRANK AND GUDGEON PINS
## STOCK NEEDLE ROLLERS

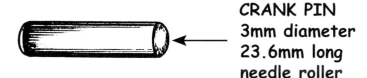

**CRANK PIN**
**3mm diameter**
**23.6mm long**
**needle roller**

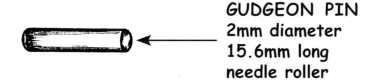

**GUDGEON PIN**
**2mm diameter**
**15.6mm long**
**needle roller**

A simple way of reducing engine friction is to use hardened needles for the crank and gudgeon pins, the type that are used in needle roller bearings.

• Discarded needle roller bearings can be a valuable source of pins for small Hot Air Engines.

• Bearing retailers are often another source of individual needles at a reasonable cost.

• Press Tool or Plastic Mould supply houses often stock a wide range of small pins.

Plastic Mould ejector pins are very handy if you are making small Hot Air Engines. Ejector pins are usually very hard, but the hard surface of the pin should be smoothed further. Slightly larger diameter silver steel rod may be used for pins, but this must be well polished.

## PLASTIC MOULD EJECTOR PINS...USES

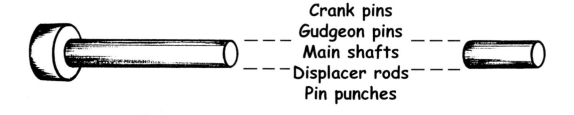

Crank pins
Gudgeon pins
Main shafts
Displacer rods
Pin punches

# CRANK-DISC AND FLYWHEELS
Aluminium, Brass, Steel

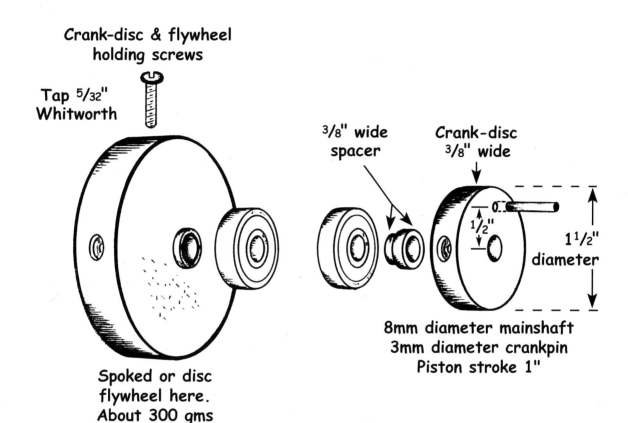

Crank-disc & flywheel
holding screws

Tap ⁵/₃₂"
Whitworth

³/₈" wide
spacer

Crank-disc
³/₈" wide

1/2"

1¹/₂"
diameter

8mm diameter mainshaft
3mm diameter crankpin
Piston stroke 1"

Spoked or disc
flywheel here.
About 300 gms

- The crankshaft and crankpin holes should be made parallel to each other if tight spots are to be avoided when the engine is turned.

- The crankpin should be a knock-in fit into its hole in the crank-disc. Drill some test holes in a piece of scrap metal first before you drill out the crank-disc for the pin.

- Hollow punch crimp around a loose crankpin or *Loctite* the pin into its hole. Drill open holes for crankpins so that they can easily be removed if necessary.

- If plain Silver Steel pins are used in the engine they should be larger in diameter to prevent them from bending. Untreated Silver Steel is rather soft.

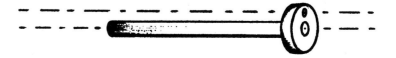

# FLYWHEELS FOR THE ENGINE
Hand Wheels, Valve Handles, Crank-Discs

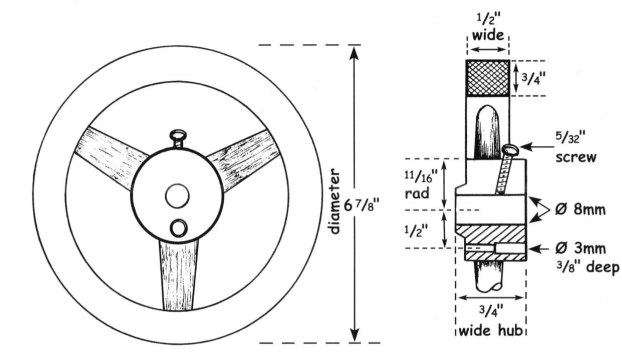

**Prototype flywheel**
**Cast aluminium**
**300gms or about 3/4 lb**

The stroke of the Engine 's piston is 1″.

The 8mm crankshaft rod should be a neat or push fit in the flywheel.

The 3mm crankpin is a knock-in fit into the flywheel.

- The flywheel needs only to be large enough to easily take the engine over a few compressions during starting. Make sure that the holding screw in the flywheel clears the connecting rod as it turns.

- The outer surfaces of the flywheel should run true to the central hole.

- The crankshaft and the crankpin holes should be made parallel to each other if tight spots are to be avoided when the engine is turning.

- It is a good idea to always drill a smaller knock-out hole so that the crankpin can easily be removed if necessary.

## ENGINE CRANKSHAFT
8mm Ground Silver Steel Rod

The engine crankshaft is made from a short length of ground 8mm Silver Steel rod which is usually a hand push fit into the ball race centres.  This stock is ideal for the job.

## ENGINE CRANKSHAFT OR MAINSHAFT
### 8mm GROUND SILVER STEEL ROD

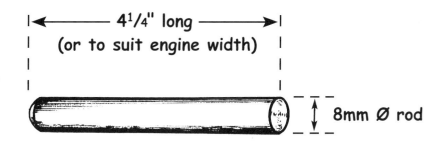

## CRANKSHAFT RETAINING COLLAR
### Brass, Aluminium, Steel

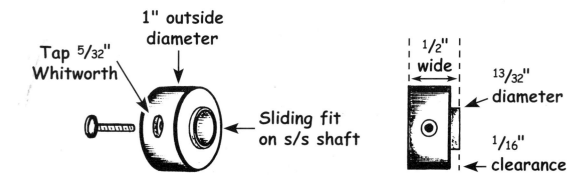

- Face both ends of the crankshaft in the lathe and round the corners with a fine cut flat file fitted with a file handle. Rounding the corners at both ends of the crankshaft helps when you are fitting the bearings.

- If you make the retaining collar for the crankshaft of a tough metal, make the clamping screw thread less than a full thread; your small taps will last longer. Try it.

- There must be NO RUBBING between the moving parts on the crankshaft and the stationary parts of the engine. *A simple, tight-fitting 'O' ring can be used as an alternative to the Crankshaft Retaining Collar.*

# CRANKSHAFT MAIN BEARINGS
2 Ball Bearing Races

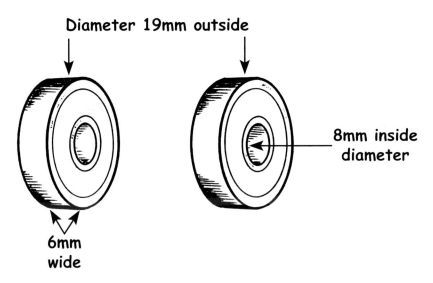

Diameter 19mm outside

8mm inside diameter

6mm wide

- Use semi sealed ball bearing races if possible. The bearings can often be cleaned of grease they may contain, or foreign matter that may have entered them.

- Use only light oil if any is needed to lubricate the ball races unless, of course, the engine is doing a lot of hard running.

- The general rule when fitting ball races is:

  *The moving part of the ball race has the tight fit (in this case onto the crankshaft rod).*

  *The stationary outer part of the ball race fitted to the bearing housing should be a neat sliding fit.*

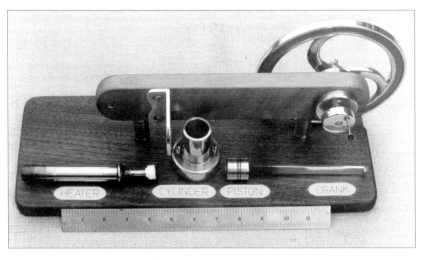

Components of Stirling 1

# BEARING HOUSINGS
## 2 required - Aluminium or Mild Steel

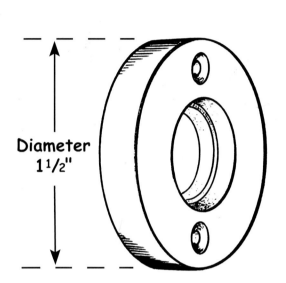

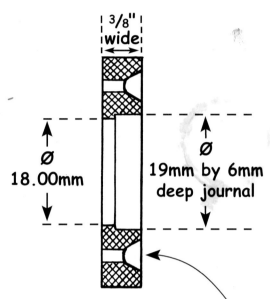

**Drill 2 clearance holes for
5/32" holding screws.
1 3/32" diameter hole centres.
(35/64" radius C/L)**

If possible run the crankshaft in semi-sealed, lightly oiled ball bearing races. Doing this helps to keep the friction in the engine low.

Remember the general rule when fitting ball races:

> **The moving part has the tight fit (in this case onto the crankshaft rod).**

> **The stationary, or outer, part (here fitted to the bearing housing) is a neat sliding fit.**

- Use a correctly sharpened cutting tool when finishing the bearing journals to size in the housings.

- Take very light cuts with the tool as you near the finished size. Clean the machined surface before you test the fit of the bearing outer in its journal.

- Mount the ball race on the end of the crankshaft rod; this is easier than holding the ball race in your fingers when testing the fit of the bearing journal.

# CYLINDER UNIT HOLDING BRACKET
Aluminium, Brass, Steel

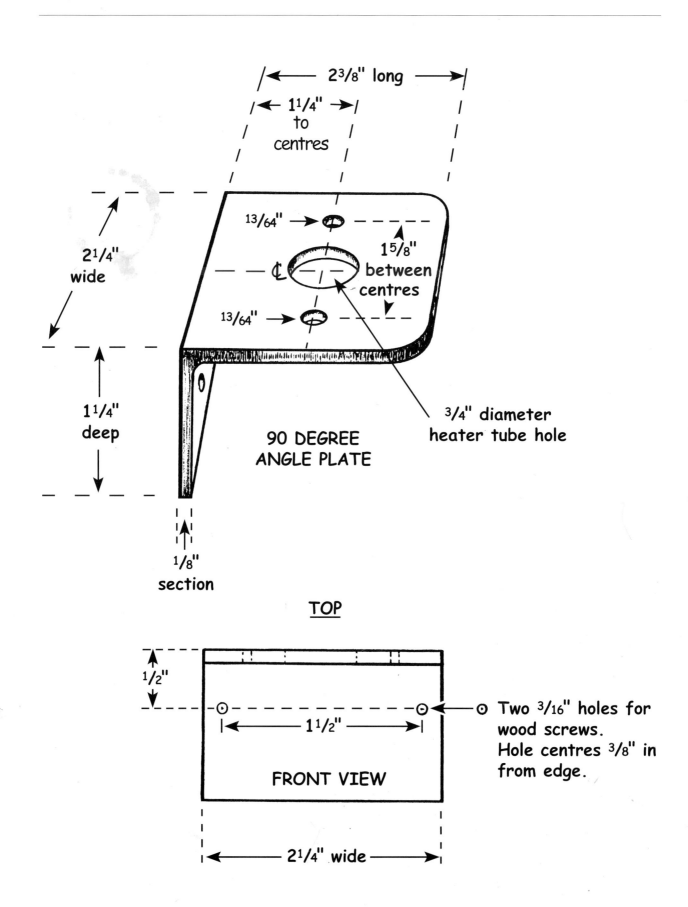

90 DEGREE
ANGLE PLATE

TOP

FRONT VIEW

Two 3/16" holes for
wood screws.
Hole centres 3/8" in
from edge.

# ENGINE CHASSIS AND BASE BOARD

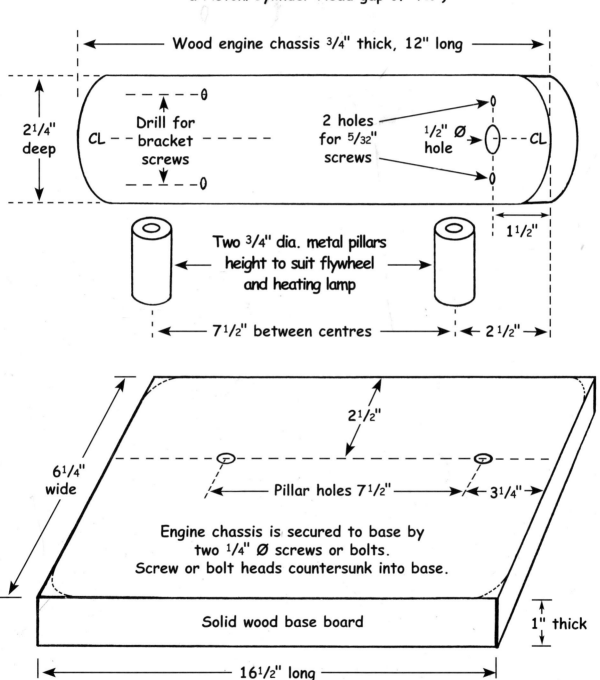

(Drill cylinder holding bracket holes to suit
a Piston/Cylinder Head gap of 1/16")

Wood engine chassis 3/4" thick, 12" long

2¹/4"
deep

CL -- Drill for bracket screws

θ

0

2 holes for 5/32" screws

¹/2" Ø hole

CL

1¹/2"

Two 3/4" dia. metal pillars
height to suit flywheel
and heating lamp

7¹/2" between centres

2¹/2"

2¹/2"

6¹/4"
wide

Pillar holes 7¹/2"

3¹/4"

Engine chassis is secured to base by
two ¹/4" Ø screws or bolts.
Screw or bolt heads countersunk into base.

Solid wood base board

1" thick

16¹/2" long

- To line-up the two bearing housings on the wooden engine chassis, drill a neat fitting hole for the mainshaft first, then secure the housings to the chassis with the mainshaft in place. Remove the parts from the chassis, drill the clearance mainshaft hole, resecure the bearing housings to the wooden chassis.

## SPIRIT LAMP
Round Aluminium Stock

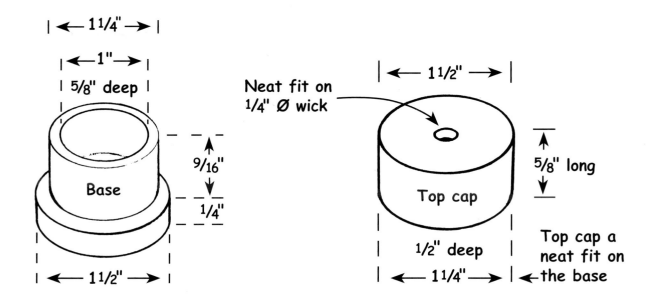

If built properly, *Stirling 1* requires very little heat to run. A suggested design of spirit lamp is shown here, but a low gas flame can be used as an alternative. *Stirling 1* is unbalanced as designed, and will bounce if run fast, so that providing some form of securing the burner is advisable once the optimum position for it has been found. However, fast running of this engine is not encouraged, so there should be little problem with vibration if a heavy base is used.

**Stirling 1 Spirit Lamp Components**

# FROM STEAM TO *STIRLING 1*

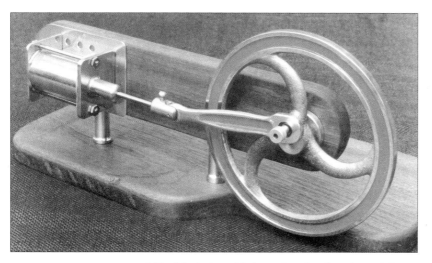

**A Double-Acting, side heated engine, based on Stirling 1.
This engine has lower R.P.M. but increased torque**

During the 1960s, the author believed that a very simple type of steam engine could be made; however, the idea was shelved because the age of steam in New Zealand was ending.

During the 1990s, Stirling Engines were added to the author's interest in old machines. Whilst making the usual two piston 'Stirlings', the author also ran a prototype of *STIRLING 1* as a simple, closed cycle steam engine.

To the keen experimenter *STIRLING 1* is really heat engines all over again, considering the number of different engines that could be made based on this simple idea. It is believed that the thermal cycle of *STIRLING 1* is similar to that of a two piston 'Stirling',

however, when the entire heater unit of *STIRLING 1* is made red hot, the thermal cycle of the engine is changed, the waste heat being dumped into the cylinder unit of the engine.

The author has, up to the present time, made dozens of small Hot Air Engines and has come to believe that almost any sort of engine will run if only the moving parts are made lightweight and very free moving.

The advice to engine builders is:

- Keep your engine designs simple; extra baggage means added friction and friction cuts the power of every engine.

- Remember to polish any plain bearing surfaces. THIS IS ESSENTIAL. Re-polish all the moving, plain bearing surfaces of your engine from time to time and your engine should continue to run well.

# FOR THE EXPERIMENTERS

- If you are making *Stirling 1* for the first time, some dimensional changes could be made without seriously affecting the running of the engine. For example: if the engine is made very free moving it will run without a flywheel which means that there is the chance that your engine would run well with a smaller flywheel than shown in the plans.

- When making the engine's heater unit, however, it would be wise to follow the instructions fairly closely.

- The *Stirling 1* heater unit was used to power a small diaphragm engine, the flexible piston being cut from a thin

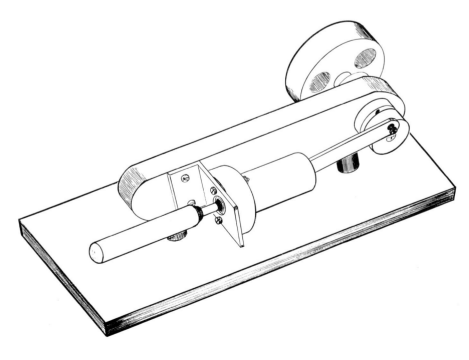

wall plastic bag. While this type of engine is simpler to make, the R.P.M. is rather low.

- Those of you wanting to make a high-speed engine, don't skimp on the ball races. An anti-friction coated piston working in a hardened steel cylinder would be in order for such an engine too.
AND KEEP YOUR HEAD WELL DOWN!

- Don't expect a double-acting hot air engine based on *STIRLING 1* to be a racer. Double-action has a damping effect on the R.P.M. An experimental double-acting engine required more heating, ran more slowly but produced more torque than a single-action engine of a similar size.

- Stainless Steel pot cleaners are a good regenerator material to use in hot air engines; however, MAKE SURE THAT THE HEATER AND ALL ITS PARTS ARE WELL CLEANED before the heater is used with an engine, or you may gum things up.

- It is possible to make *Stirling 1* without the central tube in the heater unit. Omitting this, and also altering the length of the whole heater assembly are a fruitful area for experimenting further.

# MONO PISTON HEAT ENGINES
## Some Advanced Designs

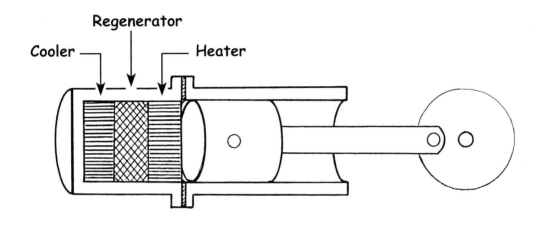

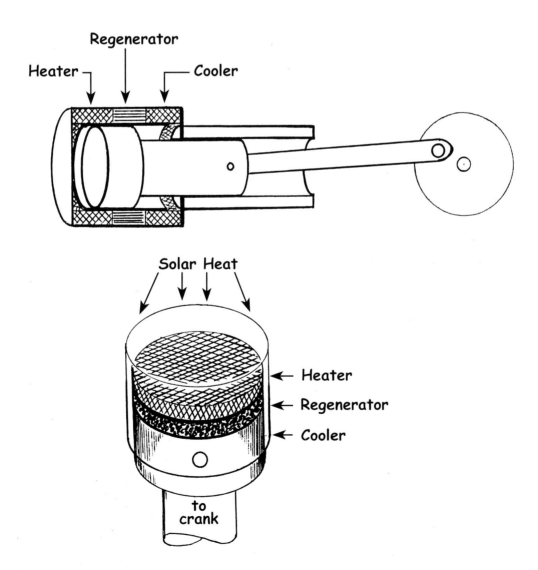

# *STIRLING 1*
## AND VARIATIONS ON A THEME

The original *Stirling 1* (above) and three other examples showing various possible configurations possible by varying the position of the heater unit.

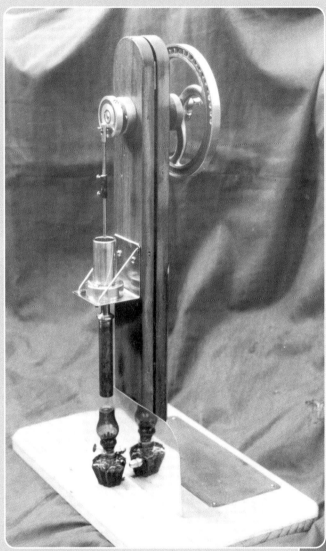

(Below) An L.P.G. Up-heated one piston engine, again with no centre tube in the heater unit.

(Above) An experimental variable stroke engine of 1″ bore and stroke. This engine is L.P.G. Up-heated and has no central tube in the heater unit.

## NOTES:

# NOTES: